DATE DUE

UNIVERSITY SCHOOL LIBRARY
UNIVERSITY OF WYOMING

LIFE ON ICE

LIFE ON ICE

HOW ANIMALS SURVIVE IN THE ARCTIC

by SEYMOUR SIMON
illustrated by SUSAN BONNERS

Franklin Watts
New York / London / 1976

*With best wishes
to Equinox Travel*

Library of Congress Cataloging in Publication Data

Simon, Seymour.
    Life on ice.

    Bibliography: p.
    Includes index.
    SUMMARY: Describes how various Arctic animals survive despite the extreme weather conditions and the threats posed by man.
    1. Zoology — Arctic regions — Juvenile literature. 2. Adaptation (Biology) — Juvenile literature. [1. Zoology — Arctic regions. 2. Adaptation (Biology)] I. Bonners, Susan. II. Title.
QL105.S55   591.5'26   75-37512
ISBN 0-531-01133-X

Text copyright © 1976 by Seymour Simon
Illustrations copyright © 1976 by Franklin Watts
All rights reserved
Printed in the United States of America
5 4 3 2 1

CONTENTS

1 Ice-Covered Sea, Snow-Covered Land   *1*

2 The Great White Hunter of the Polar Seas   *5*

3 Life in the Arctic Seas   *10*

4 Caribou and Musk Ox   *17*

5 The March of the Lemmings   *24*

6 Bird Visitors and Residents   *31*

7 Life Against Cold   *40*

8 The Circle of the Arctic Seasons   *47*

9 Man in the Arctic   *57*

Books for Reading and Research   *62*

Index   *63*

1 ICE-COVERED SEA, SNOW-COVERED LAND

Not many years ago, the North Pole was the goal of many valiant explorers. They braved below-freezing temperatures. They faced days when the winds howled at gale forces hour after hour. Blinding snowstorms made any kind of travel impossible. Time after time the violent unpredictable climate of the Arctic blocked assaults on the pole.

The American explorer Robert Peary made reaching the North Pole the purpose of his life. For twenty years, he made one attempt after another to achieve this goal. He was defeated again and again. The goal seemed almost beyond his reach when he decided to try one last time. He knew that "it was win this time or be forever defeated."

In the summer of 1908, Peary, along with his companions, started his journey over the frozen Arctic Ocean. This time he was favored by good weather. In the spring of 1909, Robert Peary, his black aide Matthew Henson, and four Eskimos finally reached latitude 90° north, the North Pole. They had reached the only point on the earth where every direction is south.

ESKIMO SLED DOGS

Today, thousands of people cross the North Pole each year. You could too, easily. All you have to do is book a seat on a commercial airline that follows the polar route between North America and Europe. Each day, many planes fly over the point that Peary found so difficult to reach.

Yet far below the warm safety of the airplane, the cold, frozen ice is as hostile to human life as it was in Peary's time. The winds still blow and the storms still rage. The animals of the Arctic still struggle to survive.

What do we mean by the Arctic? The boundary that geographers use is the Arctic Circle, the parallel of latitude at 66° 33′ north. During the first day of winter, usually December 21, the sun never comes up above the horizon at the Arctic Circle.

But the boundary of the Arctic Circle has no meaning for the animals and plants of the far north. The circle crosses the maps that humans make; the animals know nothing of it. They cross and recross the line in response to the climate and their own needs.

For our purposes, a better definition of the Arctic is the northern area of Earth where trees cannot grow. This boundary, called the tree line, wanders north and south of the Arctic Circle. Its location depends not only upon latitude, but also upon altitude, wind force and direction, summer and winter temperature extremes, and other local conditions.

Even the tree line is not an exact boundary. You won't find a forest of trees growing up to a line and then stopping. The trees just begin to thin out as you go farther north. Sometimes the trees just become fewer and fewer for as much as one hundred miles.

The trees at the boundary of the Arctic are probably not

like the ones you see in your neighborhood. They are stunted by the cold winds and short growing seasons. Some are just a few feet tall even though they may be hundreds of years old. Their trunks are thick but they narrow rapidly to the tips of their branches. Their shapes are often twisted and gnarled.

Along the protected sides of small valleys grow low-creeping shrubs such as bearberry, crowberry, and cranberry. Here and there in a protected spot you may find a small willow tree or a dwarf birch. But if you continue moving north, even these few trees disappear. The land stretches to the horizon in a rolling, treeless plain called the Arctic tundra.

The tundra may not be covered by ice during all of the year. But the largest area north of the tree line is always iced over. This is called the Arctic ice raft. It is the frozen surface of the Arctic Ocean. This enormous chunk of frozen water is one thousand miles long and six hundred miles wide. In some spots

ARCTIC WILDFLOWERS

the ice raft is hundreds of feet thick, while it is just a few feet thick in other spots.

Not all of the ice raft froze at the same time. Some parts of the ice were frozen thousands of years ago, while other parts are still forming each year. The ice raft slowly changes. Snow falls and adds to the top, while some ice melts in the sea below.

The ice is always in motion. Holes in the ice open and close. The edges break off in huge chunks that drift away. Months later, some of the chunks are seen as the icebergs that threaten ships in the North Atlantic. The whole huge ice island turns slowly from east to west, while parts of the ice move in different directions.

As you might imagine, temperatures are very low during the winter. In many places in the Arctic 50° below zero is very common and colder temperatures are often recorded. But things warm up in the summer. During June and July, the ice melts in many spots on the tundra. Water pools form and small plants grow. For short periods, colorful flowers bloom. Large numbers of insects appear. Life seems to explode furiously during the short warm spell.

Besides the summer insects, the Arctic is home to many other animals. Musk oxen, polar bears, arctic foxes, caribou, hares, lemmings, seals, and walruses are just some of the year-round inhabitants. Summer visitors include an enormous population of many kinds of birds, such as arctic terns, gulls of all kinds, and many species of swans, geese, ducks, owls, ravens, and others.

These are some of the animals of the ice-covered seas and the Arctic tundra. How they relate to each other and with their environment is the story of this book.

2 THE GREAT WHITE HUNTER OF THE POLAR SEAS

It is late March. Even in the northern reaches of the Arctic the temperature has begun to get milder. The snow layers along the northeast coast of Greenland are still many feet thick. The sun, which has not been seen for several months, has made its way above the horizon. Its long, slanting rays break through the thick clouds for a moment and turn the surface of the ice into a glistening mirror.

In the bear den many feet below the drifted surface of the snow, a mother polar bear and her two small cubs are beginning to move about restlessly. The rising temperatures of the approaching spring have roused the family from its winter sleep. The young cubs are only about two months old. If they had left their protective snow cover earlier they would have frozen to death.

The mother bear begins to dig through the layers of snow above the den. When she breaks through to the surface, she begins to travel toward the drift ice on the nearby sea. Here the cubs will be safe from attacks by wolves.

The cubs have still another enemy — male polar bears. Some males do not hibernate but are always on the hunt for food. They would welcome any animal food, including bear cubs. But a mother polar is a fierce enemy when protecting her cubs.

Some Arctic explorers have related stories of what happens when a female polar bear and her cubs meet a hunting male. The female attacks at once without any warning. A terrible battle follows that often ends with the male being chased away.

In March, seals are not out on the ice as yet. Sometimes the bears scent seals in their long caves under the snow where the seals have their pups. The bear will try to dig down and get them. But more often the seals will take to the water and escape. So the bear continues its search for food.

The seal that is breeding now and that furnishes the bear with most of its food is the ringed seal. The name comes from its fur, which has a pattern of small white rings against a darker background. Ringed seals are common and widespread in the

HARBOR SEALS

Arctic waters. They are found all along the edges of the Arctic Ocean as far north as there is open water.

Each female ringed seal gives birth to its cubs alone in a cave under the snow. Other seals, such as the harp seal and the hooded seal, breed on the shore in large groups. But the ringed seal seeks shelter below the snow where hunting bears have great difficulty finding them.

The polar bear is an animal of the sea. Most of its life is spent on the floating ice of the Arctic. Mostly carnivorous (meat-eating), the polar bear is a superb hunter. A mature male polar bear may be over eight feet long from the tip of its black nose to the end of its stubby tail. It weighs over one thousand pounds. Its fur is thick and white with yellowish tinges.

A polar bear's huge clawed feet are well adapted for running along ice or snow. Its strong powerful claws can kill a seal with a single blow. Polar bears are good swimmers and can easily make their way from one piece of floating ice to another. But the bear is no match for a seal as a swimmer and rarely catches one while swimming. Rather than trying to capture seals in the water, bears hunt seals when they climb on the floating ice to rest. They catch the seals by stealthily stalking them. The bears' whitish fur allows them to blend somewhat into the ice background. Eskimos say that while it is stalking seals a polar bear will hide its black nose with a white paw to help it remain camouflaged against the ice.

Flattening itself along the ice, taking advantage of small irregularities in the surface, a polar bear creeps along until it is within striking distance of its seal prey. Then a final rush and a single clout from a huge paw and the bear settles down for its meal.

During the winter, male polar bears hunt for food in a different way. Seals must come to the surface of the water to breathe every once in a while. When the surface is frozen, the seals melt out small openings, called blowholes, for this purpose. Even during the winter darkness, a polar bear is able to find a seal blowhole by its scent. The bear uses its claws to break up the ice to form a hole big enough to put its paw through.

The bear sits down by the enlarged blowhole and waits. When the seal comes up for air, the bear kills it instantly with a blow to its head. It is now that the enormous strength of the bear comes into play. Instead of enlarging the hole so that it can drag the seal up easily, the bear just seizes the seal with its jaws and pulls it through the small opening in the ice. The terrific force crushes the seal's ribs as it is squeezed through the opening.

A wandering, male polar bear is almost always alone. When food is plentiful in a small region, a number of bears may be in the same area. They always keep some distance apart, however, and never approach each other. Fights between bears are rare as they usually go out of their way to avoid contact.

The great white hunter of the Arctic sometimes has attendants. Polar bears are often followed at a distance by the arctic fox. The snow-white fox follows in the bear's tracks and waits for it to make a kill. Then the fox moves in to feast on any scraps

that the bear leaves at the end of its meal. Some arctic foxes may follow a particular polar bear for many weeks or even months.

In some parts of the Arctic, ravens may also follow a bear to feed on its leftover scraps. A raven has a certain advantage over a fox: It can fly and keep several bears under observation to see when a hunt is successful.

The polar bear can stand up to any animal in the Arctic except one — man. Hunters with high-powered rifles firing from low-flying aircraft have destroyed the population of bears in some parts of the Arctic. By the middle 1960s, several concerned countries had passed laws that limit the number of animals that can be killed.

These laws, along with studying polar bears' migratory patterns and food habits — carried out by an international team of scientists — are aiding the bears' fight for survival. But as with so many other of earth's animals, the polar bear can be saved only if all the interested nations act together on strict conservation measures.

FEMALE POLAR BEAR WITH CUBS

3 LIFE IN THE ARCTIC SEAS

Spring is the time of year when migratory animals start gathering in the Arctic. Some can easily be seen on land and in the air. But many of the animals that migrate north in the spring live in the sea.

Fishes, such as the salmonlike arctic char and the herringlike capelin, start arriving in large numbers. Some, such as the char, come from breeding waters in lakes at the heads of small rivers. Others, such as the capelin, live only in the sea and do not enter fresh water.

In the winter, capelin remain well out to sea. But in the spring, they move into the shallower waters along the Arctic coasts. Capelin gather in enormous shoals when they begin to move toward shallower coastal waters in May. The females are laden with roe. Each female may lay about thirty thousand eggs.

Some female capelin, accompanied by a male or two, swim toward a rocky shore just ahead of an incoming wave. They settle down on the gravel, scooping out a small hollow with their bodies and tails. In a few seconds, the eggs are laid and fertilized. The fish may complete their spawning act in time to ride the same wave back out to sea, or they may take the next outgoing wave.

Capelin also spawn in large numbers just off shore in waters a few feet deep. Sometimes the sandy bottoms are so densely covered with eggs that they shine yellow. Large groups of males roam over the spawning grounds, discharging sperm into the waters and fertilizing the eggs.

The spawning capelins are eaten in large numbers by Atlantic cod, arctic char, and other fishes, as well as seals and seabirds. In Newfoundland, Greenland, Finland, and other northern countries, capelin are caught in large numbers and salted, dried, smoked, and frozen as food for human consumption.

It is no accident that the great migrations of fishes and seals all start in the month of May. They all finally depend upon the same circumstance: the sudden blooming of the microscopic plants of the sea, the PHYTOPLANKTON. These plants make up the pastures of the sea, just as the grasses make up the pastures of the land, and form the basis for all life in the sea.

Most of the tiny plants that make up the phytoplankton are called DIATOMS. Diatoms are single-celled green plants. In sunlight, the diatoms can convert nutrient chemicals and carbon dioxide in the sea into food upon which they and all the animals that feed upon them can live. The largest cells of the plant plankton are about as big as the head of a pin. The smallest cells are invisible except under a powerful microscope. Yet the plant plankton can appear in such numbers that the surface waters are colored greenish or reddish.

In winter and early spring, there are few living plant plankton in the Arctic seas. Not enough sunlight enters the waters and it is too dim for the plants to grow. During the long, cold winter large amounts of chemicals that the plants will need when they start to grow begin to build up in the waters.

The explosion of plant plankton begins in May when the

DIATOM

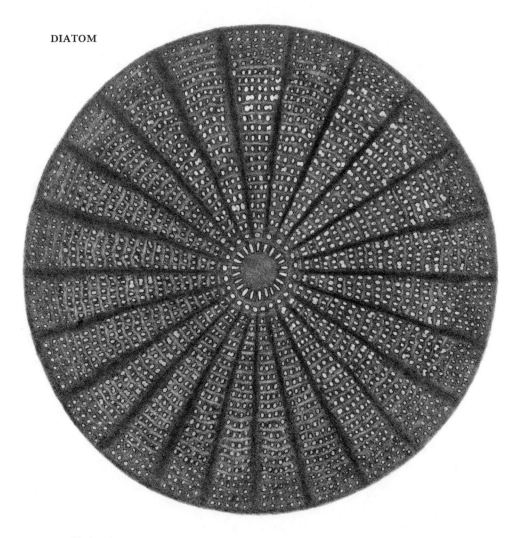

sunlight is strong and the ice cover is broken. The plants multiply furiously as if making up for their long winter sleep. At the same time, small animals of the plankton, the ZOOPLANKTON, develop in large numbers to feed upon the green, microscopic plants and on each other.

The animals of the plankton are usually a bit larger than the plants. They include many kinds of newly hatched fishes, shrimp and other crustaceans, jellyfish, sea worms, and a host of other animals. Plankton animals often have mouth parts that are able to strain small diatoms and other plants from the waters.

The plankton animals, along with the plankton plants, are eaten in turn by still larger animals. Some of these are called krill. Krill look like tiny shrimp with feathery feet. The krill are eaten by still larger animals such as fishes and sea mammals.

It is surprising that a large animal such as a baleen whale can feed directly on krill, an animal so tiny that the whale probably can't even see it. This problem is solved neatly. The whale swims in the plankton-rich sea, filtering the water through thin, horny plates of bone (called baleen) in its mouth. The krill are swallowed in enormous numbers while the smaller plant plankton just slips through. Plankton-eating seals use interlocking teeth to strain the water in the same way that whales use baleen.

The plankton-eating seals and whales are joined by large numbers of surface-feeding fishes, flocks of sea birds, and animals that prey upon all of them. The constant growing and eating in the surface waters produce a slow "rain" of living material that is used as food by the bottom-dwelling fishes, starfish, clams, and other animals of the Arctic seabed. As chief predator, humans take their choice of what to catch: fishes, whales, seals, or whatever they fancy.

The enormous growth of the plant plankton population lasts only about four to five weeks. By the beginning of June, the growing season is over in most places. All of the nutrient chemicals that accumulated over the winter have been used and plant growth stops. In some places, upwellings of cold, bottom water

BALEEN WHALE

bring nutrients to the surface, and plant growth may continue for another month or two. But in most places in the Arctic, there is little mixing of surface and bottom waters, so the plankton season is short.

All living things in the sea are part of what are called food chains. At the bottom of all the food chains are the plant plankton. Perhaps one thousand pounds of plant plankton are needed to provide enough food energy for one hundred pounds of krill. The one hundred pounds of krill furnish enough food energy for ten pounds of capelin. The ten pounds of capelin furnish

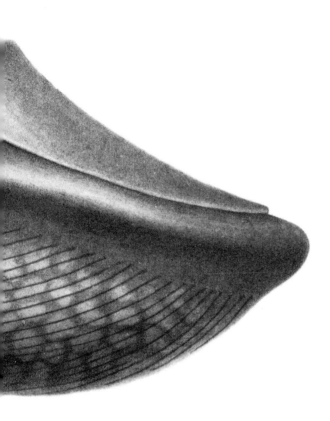

enough food energy for one pound of arctic char. In other words, for an arctic char to gain one pound in weight, one thousand pounds of plant plankton are needed at the beginning of that food chain. In each case, the remaining nine tenths of the food is used in the energy of living: hunting for food, moving about, reproducing, and so on.

Of course, there are many different kinds of food chains in the Arctic. Some end with a seal, or a polar bear, or a baleen whale. Others end with a bird, or a human being. The same kind of animal may be part of many different food chains. But each food chain starts with enormous numbers of green plants, which produce food, and ends with a few larger animals that eat smaller plants and animals.

CARIBOU

4 CARIBOU AND MUSK OX

Caribou and reindeer are really different varieties of the same species of animal, RANGIFER TARANDUS. Caribou are wild animals that live mostly in Greenland and northeastern Canada. Almost all reindeer are domestic animals and live in tended herds. Reindeer are found in northern parts of Europe and Asia as well as in Greenland.

Both caribou and reindeer are very well adapted for living in the cold regions of the Arctic. The temperature difference between their bodies and the outside air is often greater than 100°F. The heat loss from an unprotected part of their skin would be enormous. Their bodies could not produce enough heat to make up this high loss and they would freeze.

So in order to cut down on this heat loss, caribou and reindeer are covered with a thick, dense coat of hair. Each hair is a hollow tube containing air. The air spaces trap the heat very effectively. Pound for pound, a caribou's fur is probably the best insulating fur of any mammal in the world.

Beneath their dense fur covering, caribou have a thick layer of fat under their skin, which is accumulated during the fall and disappears in the spring. The fat serves as a further protection against heat loss. It is also a reservoir of stored food energy that the caribou can draw upon in winter when food is scarce.

The caribou is the only member of the deer family in which both sexes have antlers. They sweep backward and upward from the head, spreading as they rise, then bend forward. This may protect the caribou's eyes in dense forest. Their feet spread out as they walk, forming flat, broad pads that make it easy to walk on snow.

The BARREN GROUND CARIBOU is a sociable animal that lives on the Canadian tundra. It stays in herds of up to one hundred animals, except during their twice-yearly migrations. These long journeys take the caribou from their winter to their summer feeding grounds. Winters are spent in the forest zone in the company of nonmigratory caribou, called WOODLAND CARIBOU.

The food of both kinds of caribou during the winter consists of branches, reindeer moss, and grasses. Caribou sometimes use their front feet to dig through the snow cover to get at the grasses. But there is not that much available food for both kinds of caribou. The barren ground caribou uses up its store of fat and becomes very lean as spring approaches.

Beginning in April, caribou begin to mass in large herds and begin to move northward. Often the migrating herds number in the thousands. At one time in the past, the herds may have contained over a million animals. At that time, according to naturalist Ernest Thompson Seton, the caribou in one herd numbered twenty-five million!

Caribou do not follow a strict time schedule or route in their migrations. Some leave in April, others leave in May. The times and the routes apparently depend upon weather, snow cover, and other local conditions.

The calves are born in April and May and travel along with the herd. Caribou travel with a fast-moving trot that can be kept up for long periods of time. The fast-moving herds are able to outdistance wolves. Caribou are also excellent swimmers and can ford rivers and lakes with ease. Only weakened or sick animals seem to be lost to natural predators.

On the tundra, caribou feed continuously. They fatten up and begin to carry enormous reserves of food beneath their skin and in their bodies. Their antlers start to grow to large size. At this time the antlers are covered with skin through which excess heat can be shed. During the summer, the antlers begin to harden and become the weapons by which the bulls establish their dominance over each other. Caribou mate in September on the tundra and then begin their southward migration as winter begins to close in.

On its journeys and on the tundra, the caribou has few natural enemies. Some few animals (mostly calves and weak or disabled adults) are lost to the wolves, or to an occasional grizzly or a wolverine. But their most dangerous enemy by far is man.

In the past twenty years, the once-large herds of caribou have been estimated at fewer than one million. For centuries, inland Eskimos and Indians lived off the caribou herds. But the animals they took with their primitive weapons did nothing to upset the balance of nature. The herds maintained their numbers easily.

The introduction of the hunting rifle in the nineteenth century changed the balance quickly. The few difficult kills that primitive skills were able to make became a scene of massive destruction with guns. Soon the herds were destroyed and many

WOLVERINE

of the people who depended upon caribou for food faced starvation.

In recent years, protective laws and much research have slowed the destruction of the remaining herds. But whether the caribou can long survive in the changing lands of the north is still a question.

Domestic reindeer, however, are still kept and sheltered by man as they have been for hundreds of years. In fact, the earliest reference to a domestic reindeer seems to be a Chinese source dated A.D. 499. Nowadays, reindeer herds are kept by the Lapps of Scandinavia as well as other northern peoples in Europe and Asia.

Reindeer herds, in the care of herdsmen, migrate in search of winter and summer feeding grounds. Their seasonal movements are very similar to the movements of free-living caribou in Canada. Reindeer provide meat, milk, cheese, hides, and many other items for the keepers. In return, the herdsmen protect the reindeer from predators and find them good feeding grounds.

The other large, grazing animal of the Arctic tundra is the musk ox. Musk oxen are not really oxen at all. They are grouped separately by scientists somewhere between sheep, goats, and cattle. During the breeding season, the bulls give off a musky odor from facial glands, which gives rise to their name.

Looking something like a shaggy bull, the musk ox has long chocolate brown hair that falls almost to its hoofs. A thick undercovering of wool helps protect the animal against even the coldest temperatures. The musk ox has broad spreading hoofs with hair between them that helps it move on slippery ice or frozen snow.

Both male and female musk oxen have thick curved horns that cover their heads like shields. The horns curve upward to a sharp point, creating a very formidable weapon. The adult male musk ox is about four feet high at the shoulders and can weigh up to nine hundred pounds.

Musk oxen feed on grasses and other vegetation. During the winter their feeding grounds are barely enough to keep them going. They lose weight steadily as they live on their stored reserves of fat. To survive in the coldest weather, oxen form close circles huddling together for warmth.

The only natural enemies the musk ox faces in its home territories are bands of wolves. When wolves appear, musk oxen form a defensive circle with the calves and cows at the center

MUSK OX

and the bulls shoulder to shoulder around the outside. Their horns are pointing out in every direction. Few wolves will dare to attack.

Yet the very defense that proved so effective against wolves was the near undoing of the musk oxen. When human hunters found a herd, they released their sled dogs. This caused the musk oxen to form a defensive ring. From a safe distance, the hunters could shoot the entire herd without difficulty.

By the beginning of the twentieth century, musk oxen were eliminated from many areas where they had formerly lived. Only a few survived in the Arctic. Finally, in 1917, they were completely protected by law. They may have been saved just in time. There are just a few thousand left in all of the Arctic lands.

5 THE MARCH OF THE LEMMINGS

A lemming is a small, short-tailed rodent that looks like a hamster or well-fed mouse. It has a short life span of one to three years. Some years there don't seem to be very many lemmings feasting on the tundra vegetation. But other years the tundra seems to be alive with the scurrying little animals. And during the years when there are great numbers of lemmings, something very peculiar happens.

The eyewitness reports of these peak population years are very similar. It seems that during these so-called lemming years, the animal's numbers increase enormously. The lemmings seem to become very irritable and move around a great deal. Soon they begin to move across the land in living streams.

Following river valleys and moving downhill, each day the streams of lemmings are joined by more recruits that swell their ranks. Even new areas in which they can feed don't seem to tempt the lemmings to stop for long. They ford rivers and streams in their headlong journey down to the sea.

When the lemmings reach the sea, they mill around in enormous groups. As new arrivals keep piling up behind them, the lemmings plunge into the sea and start swimming. But they cannot swim across the wide waters. Thousands upon thousands die by drowning.

During their mass migration, many thousands of lemmings become food for hordes of predators. Snowy owls, foxes, weasels, bears, wolverines, hawks, gulls, jaegers, and fish feast on the lemmings. Even the caribou are seen to eat lemmings, crushing them with their teeth and swallowing them whole.

During these migrations, many other lemmings seem to die without being attacked. The dying lemmings seem to be having convulsions. Their dead bodies are found everywhere: in their burrows, around rivers and lakes, along the trail.

In a few short weeks, the lemmings are gone. Only a few individuals remain to carry on the population. Some scientists estimate that only two or three lemmings are left from every thousand. Yet these few are enough to begin the cycle again. In three or four years' time, their numbers will increase to the point where the same drama will take place.

Meanwhile, the abundance of lemmings has changed the whole balance of life in the area. The oversupply of food has resulted in an oversupply of predators, owls, foxes, and others. But the predators soon begin to suffer from a lack of food. With the arrival of cold fall weather, there is nothing for them to eat.

Many predators, particularly the snowy owl, begin to make trips to more southerly places to hunt for food. Most never find their way back north, as their numbers continue to decline.

Trappers also notice an abundant harvest of foxes in the winter just after the lemmings. With the shortage of available food, large numbers of foxes are more easily attracted to baited traps. After a lemming year, the number of foxes that are caught will decline each year until the lemmings again increase in population. The lemming years are one of the most interesting

LEMMINGS

demonstrations of how closely predators and prey fit together in the balance of nature.

Do the lemmings really commit suicide by plunging into the sea and swimming out to drown? It's very unlikely. They probably respond to the sea as just another body of water in their path. Lemmings are good swimmers and can easily ford small lakes and rivers. But the distances in the sea are just too great. The animals die of exhaustion.

But perhaps more puzzling than the reason lemmings plunge into the sea is the mystery of why lemmings increase so radically in population every three or four years. Lemmings are not the only Arctic animals to show three-to-four-year popula-

CANADIAN LYNX

tion cycles. Voles, small micelike rodents, show a similar population cycle. And we have already seen that many lemming predators, such as arctic foxes, snowy owls, and jaegers, increase on a similar three-to-four-year population cycle.

There are other Arctic animals that show population cycles that last for about ten years. Snowshoe hares, white in the winter and dark brown in the summer, have population explosions every ten years on the average. Some of their predators, such as the Canadian lynx, have the same ten-year cycle.

No one is exactly sure why there are population cycles. A number of scientists have proposed theories to explain them, but none has yet been proved. Some of the reasons are obvious.

SNOWY OWL

Large populations of grass-eating animals, such as lemmings and hares, will result in less food for every individual animal. By the same token, the varying populations of these animals undoubtedly cause the variations in their predators, such as foxes and lynxes.

Other animals are also affected by these changes. There may be less vegetation for them to eat during a lemming year. The predators will turn to other animals for food when the lemming population declines.

Weather variations may have an important effect on animal populations. Tundra plants vary from year to year in quality and quantity. A longer growing season with plenty of sunlight may be the trigger for the rapid reproduction of small rodents. On the other hand, a long, hard winter may kill off many plants and food will be scarce. In these years, rodents have fewer offspring and few of these survive.

Two good growing seasons in a row may be all that is needed. Where there were just a few rodents there are now hundreds. The great supply of food animals starts the cycle of predator increase. Then the plant food supply gives out, the lemmings start to march, the population crashes, and the cycle begins again.

Could these population cycles help the animals to survive in the cold north? Some scientists think that may be the case. Certain plant nutrients are very scarce in the northern tundras. These nurients are usually found in plants, but when the plants are eaten by lemmings, voles, and hares, the nutrients are now locked up in the animals' bodies. The nutrients again become available in the animals' droppings or in the droppings of the predators that eat the animals. The animal population cycle may be tied in with the plant nutrient needs.

With the frequently changing climate conditions of the north, the lemmings and hares could adapt quickly because there would be so many different individuals available every few years. Only the best-adapted lemmings or hares survive during the die-off period of each cycle.

Still another reason for the rapid increase and decline may have to do with the social adaptations of lemmings. Lemmings seem to breed more when their total numbers are increasing rather than stable. In fact, it has been noticed that many of the migrating lemmings are young males. Perhaps they have been forced out by social pressures of older males during breeding. In any event, the reason for the regular population cycles of many Arctic animals is still one of the great mysteries of the north.

6 BIRD VISITORS AND RESIDENTS

An Arctic tern is a gull with slender pointed white wings and tail, black cap, and orange bill. At home over the cold waters of the ocean, it darts over the surface of the water looking for plankton and small fishes. At times, a tern dips below the surface for an instant in its search for food. Its flight is swift and graceful. It's a beautiful bird to watch, but hardly very unusual, you might think.

Yet this seemingly delicate bird makes one of the longest migratory journeys known among any kind of animal, a twice-yearly trip that takes it around the entire sphere of the earth. No other animal travels as far and as often as the tern.

During the Arctic summer, terns nest on the tundra, in marshes, or on sea islands. They are found north of the tree line in Canada, Alaska, Greenland, Iceland, Arctic Eurasia, and on many of the polar islands.

Arriving in the Arctic in early June, terns breed in small colonies and lay their eggs on dry vegetation. The eggs are very well camouflaged and are difficult to see even when you are looking straight at them. The eggs are also protected by the birds of the colony, who will fiercely attack any intruder.

By August, the young fledglings and the adults are concentrating on feeding. A thick layer of fat is building up beneath

ARCTIC TERNS

their skin as provision for their upcoming journey. By early September, most of the terns have departed, leaving in small groups of ten to twenty birds.

For the Arctic terns that nested in Canada and Greenland, the first leg of their journey is across the Atlantic to the coast of Europe. During their transatlantic flight, the terns do not stop to feed or rest. Flying less than one hundred feet above the surface of the water, they keep moving in a straight easterly or southeasterly direction.

Arriving off the coast of Europe, the terns may stay for a few weeks to feed and recover from their flight. But by the end of October they are on the wing again. Now their route takes them south along the western coast of Africa. Some of the terns may winter in the cold ocean currents along the African coast, which are rich in plankton. But most continue their journey to South African seas. A few of the terns now turn northward around the Cape of Good Hope and head up into the Indian Ocean, where they may reach as far as Madagascar.

Scientists sometimes place small strips of metal around the legs of birds in an attempt to check on their movements. The strips contains information about when and where the birds were banded. A few years ago a banded Arctic tern was recovered on the east coast of South Africa on October 30. It had been banded soon after hatching on July 8 of the same year in Greenland. The young tern had flown more than eleven thousand miles in the few months since it had first begun to fly.

For most of the Arctic terns, the long journey does not end in South Africa. They continue flying still farther southward over some of the stormiest and roughest seas in the world. They do not stop on this final leg of their flight until they reach the seas around Antarctica. Here they spend the Antarctic summer (it is winter in the Arctic now), feeding in the clearings between the pack ice floating around the continent.

To get to the summer sun of Antarctica from the summer sun of the Arctic, the tern must fly at least thirteen thousand miles. And it makes this journey twice each year, a longer voyage than that of any other animal in the world.

While the tern makes the most spectacular trip, it is by no means the only Arctic bird that travels long distances. The American golden plover makes almost as long a trip as the tern. The plover is a wading or shore bird. It nests in the summer in the vegetation of the tundra. Wading in shallow water, plovers feed on small fishes, shellfish, and insect larvae.

GOLDEN PLOVER

In the late summer, plovers of the Canadian Arctic begin to move eastward and southward. By September, most have arrived on the Atlantic coast, usually in Nova Scotia. Here they fly out over the ocean, heading southward. They travel without stopping for over forty-eight hours, on a journey of over two thousand miles directly to the northern edge of South America. From there they continue to the plains of southern Brazil and Argentina, where they will spend the warm months in the Southern Hemisphere.

In the spring, the plovers return to the Arctic along a slightly different route across the middle of the continent. The whole migration forms a large oval thousands of miles in length.

There are many other bird visitors each summer in the Arctic. These include both water and land birds. All kinds of ducks, including mallards, pintails, teals, gadwalls, and widgeons, breed during the summer on the tundra marshes and wetlands. Here they build their nests in marshy, low ground and search for insect larvae in the shallow waters.

Swans and geese, such as the whistling swans, snow geese, brants, and many others, arrive in huge flocks as soon as the thaw begins in the spring. The long hours of summer daylight allow them to feed and carry on the business of breeding and raising their young.

The geese are spectacular arrivals, coming in long V-shaped waves of thousands of honking individuals. Never far from the sea, brant geese nest along tidal shores and estuaries, and the wet coastal plains. They feed on the new grasses and shoots that the warm weather brings. Other kinds of geese nest farther inshore on cliffs and rocky outcroppings that offer some protection against predators for their eggs and young.

WHISTLING SWANS

Whistling swans and Bewick's swans come to the lakes and ponds of the tundra. Here they breed and feed upon the weeds that grow along the bottoms and edges of the shallow waters. Each different kind of habitat has birds that have adapted to its special conditions.

Along the edges of the Arctic seas, other kinds of birds have arrived from near and far. Some birds are present in such enormous numbers that the cliffs and rocks of their nesting grounds are covered by thick layers of whitish droppings called GUANO.

Some shore birds such as the petrels feed at sea all year round. Gull-shaped, they glide over the water feeding on small fishes and crustaceans that they catch by dipping their heads

below the surface. Petrels usually nest on ledges along the sea. They return to them in the spring when the winter ice begins to break up.

Gulls are very social birds, feeding and living together in large flocks. They will eat almost anything from ocean plankton to fresh-water fish to the garbage that man throws away. When one food source is used up, gulls spread out searching for another source. They gather together swiftly when more food is found.

Skuas and jaegers are predatory birds that feed on shore when breeding during the summer but at sea the rest of the year. The largest species, the great skua, catch fish, attack and kill small birds, and steal food from incoming gulls. During the winter,

JAEGER

skuas and jaegers may travel far to the south. Some have even been spotted in the coastal waters of South America.

Only a few kinds of birds spend the entire year on the ice-covered lands of the tundra. These include snow buntings, rock ptarmigan, willow ptarmigan, ravens, and snowy owls. Snow buntings move south during the colder winter months, but stay within the Arctic. These small birds feed like sparrows upon grass seeds during the winter and upon insects during the summer.

Rock and willow ptarmigan are speckled brown in summer and white in winter. This change in coloring allows them to blend in with the background and keeps them from being easily spotted by predators. These pigeon-sized birds feed on the ground, digging into the thawed soil of the tundra with their claws to

SNOW BUNTING

search for insects, seeds, and tender sprouts.

Ravens are large crows that feed upon almost any type of organic material, including dead animals of all kinds, berries, and anything else that they can scavenge. They can survive in even the coldest weather. They play a useful role in the Arctic ecology, disposing of wastes and recycling needed materials into the stream of life.

Snowy owls (see page 25) are very dependent upon their principal food source — lemmings. They breed in northern Canada, Greenland, and northern Eurasia and will spend most of their time there. But their movements are controlled by the movements of the lemmings. When food is scarce, they will roam in search of lemmings, voles, or hares. In winter, they also feed upon ptarmigan and snow buntings.

7 LIFE AGAINST COLD

Only two kinds of land animals remain active during the Arctic winters: birds and mammals. Both of these animals are warm-blooded. A warm-blooded animal has a body temperature that is constant. Its temperature stays the same — or changes only slightly — even if the temperature of the air or water around it changes.

Cold-blooded animals, such as insects, snakes, or frogs, cannot survive during the Arctic winter without shelter. Their body temperature changes when the temperature of the surrounding air or the surrounding water changes. A better name for this group of animals would be variable-temperatured animals. Cold temperatures makes this group of animals so sluggish that they cannot move to find food or to escape from their enemies. They can survive only if they find shelter before the temperature drops.

Birds and mammals keep themselves warm by changing the chemical energy in the food they eat into heat energy. Of course this means that the animals must find a constant supply of food to keep themselves going. Their bodies lose heat to the surroundings just as constantly as they produce heat within their bodies.

The problem, then, is how to cut down this heat loss to the outside. And it is a serious problem. When winter temperatures drop to 50° to 60° below zero F., the difference between an Arctic animal's internal temperature and that of its surroundings may be as much as one hundred fifty degrees.

Both birds and mammals have a special way of insulating their bodies to cut down heat loss. What clothing does for man, a bird's feathers or a mammal's fur does for these animals.

The tips of a bird's feathers form a waterproof coat that protects its body from getting wet. The downy feathers close to the skin have many air spaces that trap the heat and prevent it from escaping easily.

Most Arctic mammals also have a two-layer coat, only of fur instead of feathers. The outer coat usually consists of long, oily guard hairs. This provides a protective waterproof cover for the dense woolly layer of underhair with its many air spaces. Together, the coats of hair trap and retain heat very effectively.

A thick coat of underlying fat is still another protection against the cold. Animals eat a great deal during the summer and early fall to build up the layer of fat. Seals and walruses have layers of blubber that are several inches thick to protect them against their cold environment.

Cold affects the size of animals as well as their outer coverings. Animals lose heat through their skin, so the less surface covering in relation to size the better. Thus polar animals are bigger than related animals that live in warmer surroundings.

Polar animals are also stockier in build, without projecting parts. A stocky ptarmigan is able to retain its body heat better than a slender heron. And an animal with a long tail or a projecting trunk might be very likely to have it frozen.

You can easily see how this is true by comparing the face of an arctic fox with that of a red fox that lives in a temperate climate, and with that of a kit fox that lives in a warm desert climate. The arctic fox has small ears that are mostly buried in fur. The kit fox has enormous projecting ears that radiate off a great deal of heat. But that is an advantage — in a desert. The temperate-living red fox has ears midway between the two — larger than the arctic fox's and smaller than the kit fox's.

Most Arctic mammals have short, stubby tails, short ears, and small eyes. The Arctic animals that are not like this are rarely active in the winter and usually hibernate during the colder months.

Snow itself offers a protective layer against further cold. The temperature under a thick blanket of snow may be not much below 25° above zero Fahrenheit, even though the air temperature may be fifty or sixty degrees colder than that. The reason for this is that snow, because of the many air pockets trapped within the crystals, is a good insulator. The earth has a hot core that radiates heat. Thus a thick blanket of snow traps some of the earth's heat at the surface.

This helps small mammals, such as voles and lemmings, survive during the winter. They avoid much of the cold weather by searching for food beneath the protective layer of snow. The snow also protects dormant insects, other small cold-blooded animals, and even the roots of plants from being killed off by the cold of winter.

But cold-blooded animals do have several other ways of surviving even the Arctic winter. Insects are numerous in the Arctic. But many are flightless. In adapting to the cold, they are

WALRUS

losing their wings like their cousins in the Antarctic, who do not fly at all.

The bodies of Arctic insects contain a special chemical. It is a kind of antifreeze substance. It helps to prevent them from freezing. The insects also have very little liquid in their tissues. The less liquid, the fewer the ice crystals that may form and hurt

their body cells. For example, an Arctic beetle has been found to withstand temperatures of 30° below zero Fahrenheit.

Many of these insects can freeze during cold weather and thaw out during warm weather without being the worse for wear. Scientists and explorers of the Arctic have often found frozen bodies of insects, which they would bring indoors with them. Frequently the insects would thaw and become active. One scientist even found a frozen butterfly larva, or caterpillar, that later thawed and changed into a butterfly.

Peculiarly enough, some Arctic animals are faced with the problem of getting rid of excess heat. When animals are very active in food gathering or other strenuous activities, they produce a lot of heat as a result of their exertions. Huskies, for example, could easily become overheated when they are in harness pulling sleds. Scientists have studied the problem and have come up with a partial answer, at least for how the dogs cool themselves.

For one thing, sled dogs have very few if any sweat glands on their bodies. This prevents their fur from getting wet. They are also able to flatten their hairs against their bodies and reduce the value of its insulation. When they run, their joints move and cause the fur to spread out, so that the wind can get at their bare skin and cool it off. Some heat also escapes through their furless nose and paws. Finally, dogs pant with their mouths open and their tongues hanging out. The cold air blowing against the tongue carries away some heat, and more heat is carried away by means of evaporation from the tongue.

8 THE CIRCLE OF THE ARCTIC SEASONS

As winter approaches, the Arctic day grows shorter and shorter. On the first day of winter, usually December 21, the sun never rises above the horizon at the Arctic Circle. At the circle, the Arctic night lasts for one day. But as you go farther north, the night lasts longer and longer. At the North Pole, the Arctic night lasts for six months, from the first day of autumn until the first day of spring.

The opposite happens in summer. On the first day of summer, about June 22, the sun never sets at the Arctic Circle. North of the circle, the summer day lasts longer and longer. At the North Pole, the summer day lasts for six months, from the first day of spring until the first day of autumn.

The long periods of winter night, and equally long periods of summer daylight, result from the earth being tilted on its axis at an angle of $23\frac{1}{2}°$. During the summer, the North Pole is tilted toward the sun. During the winter, the North Pole is tilted away from the sun. As the seasons pass, the tilt of the North Pole goes from one position relative to the sun to the other.

The first day of summer is called the summer solstice; the first day of winter is called the winter solstice. The word SOLSTICE comes from Latin and means "the sun standing still."

The North Pole is tipped neither toward nor away from the sun on the first day of spring and the first day of autumn. On these days, usually March 21 and September 23, the days and nights are of equal length all over the earth. These are called the vernal and autumnal equinoxes. The word EQUINOX means "equal night."

The solstices and the equinoxes mark the beginnings of each season for ASTRONOMERS. But the Arctic weather seasons are a bit more complicated than that. It would be difficult to say exactly when one season really begins and another season really ends at any one spot in the Arctic. And at different locations in the Arctic, seasons begin and end at very different times.

Still, winters and summers above the Arctic Circle are very different. Summer is brief and cool. Winter is long and bitterly cold. Spring and autumn bring important changes in the weather patterns. Living things have adapted to these conditions, and their lives change with the passing of the Arctic seasons.

Let us start our story in January in the middle of the Arctic winter. It is dark all day long. Gales and windless days alternate with each other. Snow often accompanies gales from the south and temperatures may rise a bit, but it is still many degrees below zero. The gale winds blow strong and mercilessly. Animals and men seek shelter if they can, or huddle together for protection.

Clear, windless days are colder but easier to move around in. Male polar bears continue to hunt, mostly for seals. Most female polar bears are hibernating in their dens. They will not come out until the spring and then accompanied by their newborn cubs. Seals still come up to their breathing holes in the Arctic ice and can be caught by patient Eskimo hunters warmly dressed to withstand the cold, or by polar bears.

SNOWSHOE HARE

The willow ptarmigan and the snowshoe hare are hardy animals that move about all winter long on the tundra. They have both turned white for the winter and blend in well against the snow. The ptarmigan has also grown fluffy clumps of feathers on its legs and feet. These provide insulation against the cold and pad out the bird's feet so that it can walk along the surface of the snow without sinking down.

The snowshoe hare has its own kind of snowshoes. It has grown clumps of fur on its large hind feet. Like the snowshoes worn by humans, the clumps of fur help distribute weight over a larger area of snow and help to prevent the hare from sinking into the snow.

February and March are still very cold, but the sun is climbing higher and higher and the days are getting longer. Sometimes rain rather than snow will fall. The rain then freezes into ice,

PTARMIGANS

often trapping some animals in its grip. For the grazing animals of the lower Arctic, the frozen rain may spell disaster. They cannot feed on the ice-covered grasses.

April and the beginning of spring are not what you might imagine. There are no plants budding in the upper Arctic and no spring insects. But the migrant animals have begun to arrive. Snow buntings are reaching the upper Arctic by the end of April. Some may even come before there is any food available and die. By the end of the month, the large flocks of sea birds are beginning to arrive and nest. The caribou are also on the move, leaving their southern range and starting their journey to northern feeding grounds.

Young seal pups lie about on the ice. They cannot swim until three weeks old. For seal hunters, such as the polar bear and man, this is the ideal time to hunt for these animals.

About this time of year, the ground squirrel wakes up from hibernation and begins to search for food. In the three or four months they are active, squirrels must store up enough fat to keep them alive during an eight- or nine-month-long hibernation.

Scientists have studied the ground squirrel and have found out some interesting facts about its habits. The squirrels hibernate in places that are covered by deep, soft snowdrifts during the winter months. The snow forms a protective blanket against

GROUND SQUIRREL

the colder winter temperatures, and ground temperatures hover around 10° above zero Fahrenheit.

But even that temperature would be too cold for a hibernating animal to survive the winter. The squirrels are able to keep their nests warmer than the surrounding soil in several ways. For one thing, even though their body processes slow down during hibernation, the squirrels still produce much body heat. Their nest burrows are lined with grasses, which help to keep the heat from escaping.

The design of the burrow is very unusual. The entrance is always lower than the resting chamber. Because warm air rises, it becomes trapped in the chamber rather than escaping through the entrance. The burrows are always located in well-drained soil that doesn't become waterlogged. If the soil stayed wet, the heat would quickly drain away from the resting chamber during snow melt.

May and June bring the end of winter and the beginning of summer. The snow melts on the tundra. However, it never melts in the far north. Most of the bird and other animal migrants have arrived by May. The birds have nested, laid their eggs, and hatched their young. But a sudden cold spell or spring storm may still kill many birds in these areas.

In the seas, the longer hours of daylight result in a great flowering of the plant plankton, and of the animal plankton that feed upon them. In turn, the fishes, seals, and all the other parts of the food chains link together in this time of plenty.

By the middle of May, the lower Arctic is full of flowering plants on the tundra. Insects are plentiful. The weather is warmer, but hardly what you might think of as hot. There are long days

SNOW GOOSE

of rain and sunny days are rare. In the upper Arctic, there still may be snowstorms.

Animals begin to molt, losing their winter furs or feathers, and change to a lighter summer garb. Seals pant in what is hot weather to them and plunge frequently into the cold Arctic waters to cool off.

In the upper Arctic, June is flower month. The mosquitoes of northern Canada come in swarms of fierce biters. On the coasts, the whales arrive from the south. Streams of cold water trickle down from the glaciers. The sea cliffs are alive with birds and the animal population soars.

July is high summer all over the Arctic. Plant life is plentiful and the animals feast on the plants and on each other. Birds are busy feeding their young. But some of the migrants, such as the female phalarope, have even begun to move south by the end of the month. Glaciers are breaking off huge chunks into the sea. This process, called calving, results in the formation of icebergs.

August, while still summer, begins to hint of approaching changes. The birds are almost all flying south. They have layers of fat on their bodies, which will provide energy for their long journeys. The adult birds usually leave earlier, while the young may not leave until September. Adult geese have to wait until molting is complete before they can leave. They cannot fly until they have their new feathers. August is often a hot month and the sea ice is melted more now than at any other time.

September brings a cold autumn and great changes. The long summer days are over. Days and nights are now the same

PHALAROPE

length. Frosts kill the plant life on the land. Streams and lakes freeze over rapidly. Caribou mate now. Animals, such as foxes, hares, and lemmings, start to change the color of their fur. Insects die and the Arctic begins to prepare itself for winter.

In October, the days rapidly get shorter and shorter in the low Arctic, while in the high Arctic the sun has already disappeared below the horizon until the next spring. Polar bear females begin to find dens for the winter where they can have their young. Ground squirrels burrow out their nests before the ground freezes.

By November, winter has settled in hard. Snow and ice are everywhere. The sun is gone from the sky and few animals are seen. Instead of the sun, there are often brilliant displays of colored lights in the sky, the AURORA BOREALIS. But there are few eyes to see the almost nightly auroras. Most animals are in their dens and the Eskimos spend the month visiting each other while their preserved food supplies are still full.

The end of the year brings the darkest month — December. It is not as cold as it will get in a month or two and the skies are often clear. Yet the land is lonely and desolate. Nothing seems to be alive. But life hangs on even during the cruel winter. And when the sun returns again in the spring, life is renewed and once again there is plenty.

LAPP MOTHER AND CHILDREN

9 MAN IN THE ARCTIC

For thousands of years, humans have survived in the Arctic. The Lapps of northern Scandinavia, the Samoyeds of north Russia, the Tungus and Yakuts of eastern Siberia, the Eskimos of North America and Greenland all prove that even the harsh conditions of the Arctic can be conquered by people of skill and hardiness.

With intelligence and tools the native peoples of the Arctic have overcome the handicaps of human bodies that are better suited for milder climates. With strength and determination they have endured conditions of cold, long days of hunger and fatigue, endless barrens of snow and ice, howling gales, and snowstorms.

Each of the peoples of the Arctic is different in many ways. Some are herders of reindeer. Others are hunters and fishers. Some live close to the coast and depend upon the ice-covered sea for food. Others live inland, trap animals, catch river fish, and depend upon the migrating caribou.

Yet the primitive peoples of the Arctic have much in common. The same kinds of climate conditions and the same kinds of materials available to them have forced them to use similar methods to survive. They all dress in the furs or feathers available from hunting. Their clothing includes loosely worn parkas or coats and breeches made of skins. They wear boots of reindeer or sealskin. They have fur mittens and sleep on rugs of polar bear or caribou fur.

In the lower and warmer parts of the Arctic, the native peoples use tents of animal skins stretched over driftwood or birch poles. In the high Arctic, Eskimos build igloos of ice blocks to protect themselves against the winter gales and below-freezing temperatures.

Many of the different tribes use sledges, snowshoes, or skis to get around more easily. Some use Arctic dogs such as huskies, malamutes, or Samoyeds (the name of a people and a kind of dog), to pull their sledges, while others use reindeers as pack animals.

During their long history in the Arctic, primitive peoples developed the understandings and skills that were needed for survival. They became nomadic hunters and herders. Their populations were small and remained small because of the lack of enough food and the bad weather conditions. Yet they fitted into the ecology of the Arctic without harming it or themselves.

But the impact of the more technological civilizations of the south has changed man's role in the Arctic in many ways. Commercial fisheries and fur hunters are replacing the individual. Instead of making their own weapons and clothing, many native northerners now buy them from catalogs. Instead of using a harpoon to kill seals and polar bears, they now use high-powered rifles.

Are all of these changes for the good? Isn't it right for the peoples of the Arctic to benefit from modern technological progress? Like all peoples of the world, the Eskimos and other northerners have a right to share in the benefits of scientific advances. And yet sometimes this progress can be harmful.

ESKIMO MOTHER AND CHILDREN

To pay for the guns, clothing, and food the Eskimos have begun to overhunt for furs and other valuable animal products. The numbers of animals have declined. Some kinds have become EXTINCT, others are on the verge of extinction. Once poor but able to survive on their own, some Eskimos now have to depend upon handouts for their survival. The dangers of altering the ecology of the Arctic have become quickly apparent.

We can see this in the history of the exploitation of sea animals. Once whales, walruses, and other sea mammals were plentiful in the cold waters. Some kinds of whales, such as right whales, strain plankton directly from the sea through large plates of baleen. Other sea mammals feed on the plentiful fish in the waters. For many years the whales and other mammals were hunted without limit.

Baleen, or whalebone, was used in many ways as a strong springy material. In addition, whale's blubber yielded high-quality oil for the oil lamps that were used before the days of electricity. Whale hunters killed off the right whales ruthlessly, until today there are few if any left. The hunters also destroyed many colonies of seals, huge numbers of walruses, as well as such land animals as the Hudson Bay musk oxen.

Modern whalers with their explosive harpoons and fast ships next went after faster-moving whales, such as the blue whale, the finback, the sei, and the minke. By now most of these are gone or going fast. Instead of harvesting these animals in small enough numbers so that they could maintain their stocks, hunters have destroyed them for future generations.

Fur-bearing animals have fared little better. Enormous herds of harp seals have been harvested in great numbers since 1800. In 1889, for example, whalers took three hundred thousand harp seal skins. They killed millions more that were lost in the sea and not recovered. As recently as 1964, films of the seal hunts, which were shown on television, shocked and revolted many people around the world. Viewers saw that large numbers of pups were killed along with mature seals in horribly bloody ways.

Finally the Canadian government began to regulate the number of seals that could be killed yearly and the ways in which they could be taken. But this annual quota is thought by many biologists to be too high. Now an even greater threat to sea mammal existence in the Arctic lies in the unrestricted fishing that is going on. Fishing vessels may soon be harvesting so many fishes from the Arctic seas that the sea mammals may not have enough food.

All of these activities lead to serious questions. It is true that fish are needed for human consumption. But governments must find a way to effectively regulate the harvest of these food animals without affecting all the other animals that also depend upon them. We are not the only living things in our world. And if humans destroy everything else we will not long survive.

The latest and perhaps most dangerous hazard for the Arctic is the recent discovery of oil on the north Alaska coast. Although precautions are being taken, there are many chances for disaster. A large oil spill from a tanker or a blown-out well could result in major damage to the life of the area. Much of the oil will be coming south via a trans-Canadian pipeline. This line, which goes through the tundra, may have untold effects on the ecology of the region. It is still too soon to know.

The only chance for the Arctic to continue without being completely destroyed is for intelligent planning to take place. It is inevitable that there will be changes in the Arctic. But can we make these changes without fouling the land and the sea? We can save the Arctic only if enough people insist that governments take the time and effort to care enough. The choice is up to all of us.

BOOKS FOR READING AND RESEARCH

Dyson, James L. *The World of Ice.* New York: Knopf, 1962.

Freuchen, Peter, and Salomonsen, Finn. *The Arctic Year.* New York: Putnam, 1958.

Fuller, William A., and Holmes, John C. *The Life of the Far North.* New York: McGraw-Hill, 1972.

Laycock, George. *Alaska, the Embattled Frontier.* Boston: Houghton Mifflin, 1971.

Ley, Willy, and the editors of *Life. The Poles.* New York: Time-Life, 1962.

Murie, Adolph. *A Naturalist in Alaska.* New York: Doubleday, 1963.

Perry, Richard. *The World of the Polar Bear.* Seattle: University of Washington Press, 1967.

―――. *The World of the Walrus.* New York: Taplinger, 1968.

Pruitt, William O., Jr. *Animals of the North.* New York: Harper & Row, 1967.

Snyder, L. L. *Arctic Birds of Canada.* Toronto: University of Toronto Press, 1957.

Stonehouse, Bernard. *Animals of the Arctic.* New York: Holt, Rinehart & Winston, 1971.

Sutton, George M. *High Arctic.* New York: Erikson, 1971.

INDEX

American golden plover, 34–35
Arctic Circle, 2
Arctic ice raft, 3–4
Arctic Ocean, 1, 3
Arctic tundra, 3, 4, 52
Aurora Borealis, 55
Autumn, in Arctic, 47, 54–55
Autumnal equinox, 48

Baleen, 60
Baleen whale, 13, 60
Barren ground caribou, 18
Bearberry, 3
Bears. *See* Polar bear
Beetle, 45
Birds, 4, 25, 31–39, 40–41, 52, 53–54
Blowholes, 8
Blubber, 41, 60
Body temperature, 40

Calving, of icebergs, 54
Canada, hunting regulations, 60–61
Capelin, 10–11
Caribou, 4, 17–20, 25, 55, 57
Char, 10, 11

Climate, 47–55
 and animal populations, 29–30
Clothing, 57–58
Cod, 11
Cold-blooded animals, 40, 43–46
Conservation laws, 9, 23, 59–61
Cranberry, 3
Crustaceans, 13

Diatoms, 11, 13
Dogs, 46, 58
Ducks, 4, 35

Ecology, Arctic, 39, 58–61
Equinoxes, 48
Eskimos, 19, 48, 55, 57, 59
Exploitation, of Arctic, 58–61
Extinction, of animals, 59

Fat, as protection from cold, 41
Feathers, 41, 53, 54
Fish, 10, 11, 13, 25, 52, 61
Fishermen, 11, 57, 61
Food chains, 11–15, 25–30, 52
Fox, Arctic, 4, 8–9, 25, 27, 29, 43, 55

(63)

Frogs, 40
Fur, 41, 43

Gales, 48
Geese, 4, 35, 54
Glaciers, 53, 54
Greenland, 5, 11
Guano, 36
Gulls, 4, 25, 37

Hares, 4, 27, 29, 30, 49, 55
Hays seal, 7
    See also Seals
Hawks, 25
    See also Birds
Heat, body, 40
Henson, Matthew, 1
Hibernation, 5, 43, 51, 52
Hooded seal, 7
    See also Seals
Humans, 9, 13, 23, 57–61
Hunting, 57–61
Husky, 46, 58

Ice raft, 3–4
Icebergs, 4, 54
Igloos, 58
Insects, 4, 40, 43, 44–46, 52, 55
Insulation, 41–46

Jaeger, 25, 27, 37–38
Jellyfish, 13

Krill, 13

Lapps, 20, 57
Lemming, 4, 24–26, 29, 30, 39, 43, 55

Lynx, 27, 29

Malamute, 58
Mallards, 35
Mammals, 40–43
Man, 9, 13, 23, 57–61
Migrations, 10, 18, 24–25, 31–34, 35, 52
Molting, 53, 54
Mosquitoes, 53
Musk ox, 4, 21–23, 60

North Pole, 1–2, 47

Oil, as hazard, 61
Owls, 4, 25, 27
    See also Birds

Pack animals, 58
Parkas, 57
Peary, Robert, 1, 2
Petrels, 36–37
    See also Birds
Phalarope, 54
    See also Birds
Phytoplankton, 11
Pipeline, trans-Canadian, 61
Plankton, 11–13, 52
Plant life, 3, 4, 11–13, 29, 43, 52
Plover, 34–35
Polar animals, 41
Polar bear, 4, 5–6, 7–8, 25, 48, 55
Population cycles, 25–30, 53
Predator cycles, 25–30
Ptarmigan, 38, 39, 41, 49

Ravens, 4, 9, 38, 39

Reindeer, 17, 20, 58
Right whale, 60
Ringed seal, 6–7
    See also Seals
Rock ptarmigan, 38, 39
Rodents, 24, 27, 29

Samoyed (dog), 58
Samoyeds (people), 57
Sea worms, 13
Seals, 4, 6–7, 8, 13, 41, 52, 53, 60–61
Seasons, Arctic, 47–55
Seton, Ernest Thompson, 18
Shrimp, 13
Skins, as clothing, 57–58
Skis, 58
Skuas, 37–38
Sled dogs, 46, 58
Sledges, 58
Snakes, 40
Snow, as insulator, 43
Snow bunting, 38, 39
Snowshoe hare, 27, 49
Snowshoes, 58
Snowy owl, 25, 27, 38, 39
Social animals, 18, 21, 24, 30, 37
Solstice, summer, 47
Spawning, 10–11
Spring, in Arctic, 48, 49, 50–52
Squirrels, 52, 55
Summer, in Arctic, 47, 52–54
Summer solstice, 47
Sun, 47, 49, 55
Swans, 4, 35, 36

Technology, and Arctic peoples, 58
Temperature, 4, 41
Tents, 58
Terns, 4, 31–34
Transportation, 58
Trappers, 57–58
Tree line, 2–3
Trees, 2–3
Tundra, 3, 4, 52
Tungus, 57

Vegetation, 3, 4, 11–13, 29, 43, 52
Vernal equinox, 48
Voles, 27, 43

Walruses, 4, 41, 60
Warm-blooded animals, 40
Weapons, hunting, 58
Weasels, 25
Weather, 47–55
    and animal populations, 29, 30
Whalebone, 60
Whales, 13, 53, 60
Willow ptarmigan, 38, 39, 49
Winter, in Arctic, 47–49, 51–52, 55
Wolverine, 25
Wolves, 5, 21
Woodland caribou, 18

Yakuts, 57

Zooplankton, 12–13

(65)

ABOUT THE AUTHOR

Seymour Simon is at present teaching science in a New York City school. Mr. Simon is the author of over thirty science books, including *Animals in Field and Laboratory, Exploring with a Microscope, From Shore to Ocean Floor, Projects with Air,* and *Life in the Dark,* all published by Franklin Watts.